Little Peach Pit Presents:

WHOLESOME

Together we can save the planet!

Grace Nava, Ed.S.

ISBN-10: 0967506859
ISBN-13: 978-0967506852

Library of Congress Cataloging-in-Publication Data
Media for Life Corporation, Little Falls, New Jersey

"Wholesome!"

By Grace Nava, Ed. S.

Illustrated by:
Kübra Aslan & Grace Nava

Printed in the United States of America.

Published by
Media for Life Corporation
PO Box 1214
Little Falls, NJ 07424

This book is available for special promotions and premiums.

Visit: Little-Peach-Pit.com

Reviews are deeply appreciated!

To Kairos

One morning, Little Peach Pit got up early to take a walk. It wanted to see the natural world surrounding its lovely home in the orchard.

As it went on its way, it noticed dust coming on its little feet. Others may have felt bad that their feet got dirty, but Little Peach Pit said,

"Thank you soil for giving nutrients to so many trees and plants (1). They make us flowers, fruits, and vegetables. Without a healthy you, my life would not be complete."

The soil giggled as it felt Little Peach Pit's steps.

Then Little Peach Pit felt a few raindrops. It looked up and noticed several fluffy clouds passing by. Others may have felt bad that they were getting wet, but Little Peach Pit said,

"Thank you clouds for watering the trees and plants (1) I love so much. Without a healthy you, my life would not be complete."

The clouds waved good-bye to Little Peach Pit as they moved on to water other parts of the world.

Little Peach Pit was whistling a happy tune when it was about to step on an ant. It stopped just in time! Others might have laughed at the ant or be afraid of it because it was an insect, but Little Peach Pit said,

"Good morning lovely ant. I am sorry I was about to step on you. You work very hard to keep the soil rich (2). Without a healthy you, my life would not be complete."

The ant thanked Little Peach Pit for its kindness and hurried with its heavy load to its anthill.

Little Peach Pit heard a loud buzz and quickly moved out of the way. It was a busy bee looking for flowers to drink their nectar. Wow! That was close, it almost got stung! The bee said,

"My apologies! Winter is coming and there are only a few flowers left. I must hurry!"

Others might have feared the bee because of its stinger, but Little Peach Pit replied,

"No worries! I'll stay out of your way. Thank you for working so hard. Your feet take pollen from flower to flower, pollinating them (3). Because of you, there are juicy peaches, apples, and many other delicious fruits. Without a healthy you, my life would not be complete!"

The bee flew in a circle as a sign of appreciation for Little Peach Pit's kindness and dashed to a patch of beautiful sunflowers.

Little Peach Pit reached a pond and saw a frog jumping to catch a fly. Others might have disliked the frog for being green and slimy, but Pit said,

"Good morning, frog! I hope your breakfast was tasty."

The frog replied,

"Indeed! I like flies and mosquitoes the best."

Little Peach Pit replied,

"Well, then, I am glad you have plenty of food here. Because you look happy, I know the pond is thriving with many fish and other creatures (4). Without a healthy you, my life would not be complete."

Pit showed a few mosquitoes the frog's way, who thanked it for the good gesture.

As Little Peach Pit got closer to the city, it noticed that the soil looked thin and sick. It had only a few weeds growing here and there. With a frown, it asked,

"What is wrong soil? You don't look very healthy."

The soil replied,

"People cut down too many trees and other plants. The fewer plants that live here, the fewer roots there are to hold me in place when the wind blows and the rain runs downstream (5). I might become thinner and thinner, until one day, I'll be all gone!"

Pit replied,

"Oh, My! Is there anything I can do to help?"

"Yes," replied the soil. **"Tell people to plant more trees so I don't get washed away!"**

Little Peach Pit replied,

"I will soil, because without a healthy you, my life is not complete."

Then Little Peach Pit felt a few rain drops. When it looked up, the clouds did not look happy.

"What is wrong?" it asked.

"We have acid rain! When we pass by areas where factories and cars release pollutants into the air, we pick them up and carry them to other areas, even other countries. Sometimes we kill many trees and plants because our raindrops have poison in them!" (6)

"Oh, my!" replied Little Peach Pit, and asked, **"Is there anything I can do to help?"**

"Sure!" said the clouds. **"Tell people to keep bad chemicals from the air!"**

Pit replied,

"I will, clouds, because without a healthy you, my life is not complete."

Little Peach Pit kept walking toward the city, but as it got closer to downtown, it did not meet any bees or ants, and the pond at the park looked dirty. A very sickly looking frog was trapped in a plastic bag (7). Pit quickly helped the frog free itself from it, but it was too late. The frog's legs were hurt. The frog said,

"I am so hungry but I cannot find good enough food to eat."

"That's terrible!" said Little Peach Pit. **"Is there anything I can do?"**

"Yes," said the frog. **"Tell people to throw garbage into garbage cans, not the pond, to use less plastic bags, and reuse bags and other things. There is too much garbage!"**

Pit said,

"I will, because without a healthy you, my life is not complete."

As Little Peach Pit kept walking, its little feet burned from the hot asphalt. Everywhere it looked, there was no good ground with healthy soil showing trees and plants. Rain then started to pour. The cool rain felt refreshing on its feet. However, the rain was running very fast downhill in streams, taking so much garbage toward the river (8). The little streams of water screamed,

"Help! We cannot stop."

And as they ran downhill, they started to flood the homes near the river. Pit asked,

"Is there anything I can do?"

"Yes!" said the streams. **"Tell people to leave areas unpaved where we can be absorbed by the ground, and make our way to the river slowly underground."**

"I will," said Little Peach Pit. **"Because without a healthy you, my life is not complete."**

Little Peach Pit kept walking and on its way it found people cleaning a street. Enthusiastically, it asked if it could join them. The organizer, a teenage girl, replied,

"Sure, the more the merrier! We can finish faster with more help."

The garbage was separated into recyclable, compostable, and non-recyclable.

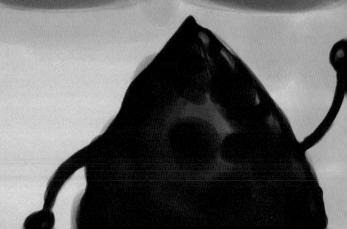

CLOTHES

PAPER   METAL   PLASTICS   GLASS

The recyclables were taken to a recycling center. People were busy sorting plastics, metals, and clothes, so they could be reused. Recycling helps to reduce the waste that ends up in landfills (9). Little Peach Pit was very encouraged that people were recycling and thought of its friend the frog who would be happy that garbage would not end up in the pond. When they were done, Pit said,

"Thanks for letting me be part of your effort!"

The people waved good-bye and said,

"Thanks for your help! Come back next month when we'll do it again!"

As Little Peach Pit started to make its way home, it noticed some people very busy at an empty lot. They had removed all the garbage and were creating a vegetable community garden where people got together to grow fruits and vegetables that the families in the group would enjoy (10). Pit thought of its friend the soil and felt happy there were going to be many roots to hold it in place. Pit joined the gardening group and got its little hands into the rich soil and happily planted tomato, zucchini, and cucumber plants As it left, the people said,

"Thanks for your help! Come back at harvest time to celebrate with us."

To which Little Peach Pit replied,

"Thank you. I will!"

Then Little Peach Pit noticed a charging station for cars. People were getting their energy to run their cars from electrical outlets. It felt happy because electric cars don't pollute as much as gas ones (11). It thought about its friends, the clouds. They would not have to carry bad things and give acid rain. While a man waited for his car to charge, he asked what had brought Pit that way. It replied,

"I am exercising, walking is very good for one's health (12). I am glad you are using an electric car, but it would be even less polluting if you and your family walked to places."

"**I like your idea,**" said the man, "**but in some places in my neighborhood, there are no sidewalks which make it unsafe for pedestrians. I will tell my city to make it safer, though. Perhaps we can designate 'pedestrian traffic only' areas.**"

"**That's great!**" said Little Peach Pit as it waved good-bye.

NO PESTICIDES

Little Peach Pit left the city. In the countryside, it found an organic farm, where no pesticides were used. The farm animals were happily feeding on grass (13). One cow said,

"Would you like to join us for a break?"

Pit said,

"Sure!"

After enjoying the coolness of the shade provided by a lush tree, Little Peach Pit continued its way home.

Now, Little Peach Pit has a message for you:

"Plant trees and plants, especially those that you can eat, so you can save money, have less to shop for, and consume local produce. Keep the air clean, recycle all you can, reuse materials in creative ways, and reduce waste. Also, remember to clean after your dog, and keep the garbage in its place."

"Search for ways to make our world a happy place for all to live. You can become a lawmaker who can create laws that protect the environment or a scientist who develops new ways for people to eat and live without hurting the Earth and its creatures." (14).

"Because without healthy soil, rain, animals, and plants, our lives cannot be complete. Big or small, everything around us plays a part in providing a wholesome environment for all to enjoy!"

Vocabulary

Absorb: Becoming part of something else

Acid Rain: Rain that has bad elements for plants and animals

Compostable: Safe things that can go into the soil to make it rich

Enthusiastically: Something done with energy and great interest

Environment: The natural world that all life must live upon

Landfills: Areas where garbage is put that cannot be composted or recycled

Non-recyclable: Garbage that cannot be composted or recycled and must go into landfills

Pedestrians: Another word for people who are out in public walking

Pollinating: Taking pollen from one plant to another to make flowers and fruits grow

Polluting: Putting bad things into the air, soil and water

Recyclable: Used things that can be remade into new things

Wholesome: When all natural parts combine that are meant to be together

Resources

Gardening

BBC Gardening with Kids
http://www.bbc.co.uk/gardening/gardening_with_children/
KidsGardening.org
http://www.kidsgardening.org
My First Garden
http://extension.illinois.edu/firstgarden/

Recycling

California Department of Conservation
http://www.energyquest.ca.gov/saving_energy/RECYCLINGFactsGamesCrafts02.PDF
Kids Recycle Project
http://kidsrecycle.org
Science Kids
http://www.sciencekids.co.nz/recycling.html

Environmental Sciences

Office of Citizens Services and Innovative Technologies
https://kids.usa.gov/jobs/a-z-list/index.shtml
National Institute of Environmental Health Sciences
http://kids.niehs.nih.gov/index.htm
North Carolina Department of Agriculture & Consumer Services
http://www.ncagr.gov/htm/educational.htm
Science Buddies
http://www.sciencebuddies.org/science-fair-projects/project_ideas.shtml#browseallprojects

References

(1) North Carolina Department of Agriculture. Plant Nutrients [Internet]. NCAGR.gov. 2016 [cited 26 February 2016]. Available from: http://www.ncagr.gov/cyber/kidswrld/plant/nutrient.htm

(2) Sanders D, van Veen F. Ecosystem engineering and predation: the multitrophic impact of two ant species. Journal of Animal Ecology. 2011;80(3):569-576.

(3) Newshour. [Internet] Are pesticides to blame for the massive bee die-off? [Internet]. PBS. 2015 [cited 26 February 2016]. Available from: http://www.pbs.org/newshour/bb/are-pesticides-to-blame-for-the-massive-bee-die-off

(4) Sciencebuddies.org. Froggy Forecasting: How Frog Health Predicts Pond Health [Internet]. 2016 [cited 26 February 2016]. Available from: http://www.sciencebuddies.org/science-fair-projects/project_ideas/EnvSci_p048.shtml

(5) Silva-Sanchez, N., A. Martinez Cortizas, and L. Lopez-Merino. Linking Forest Cover, Soil Erosion And Mire Hydrology To Late-Holocene Human Activity And Climate In NW Spain. *The Holocene* 24.6 (2014): 714-725. Web.

(6) Meteorology Manual: The Practical Guide to the Weather. What Are the Causes and Effects of Acid Rain and What Can Be Done to Tackle the Problems? Why Is Acid Rain a Global Problem?. The Hutchinson Unabridged Encyclopedia with Atlas and Weather Guide. Abington: Helicon, 2015. Credo Reference. Web. 26 Feb. 2016.

(8) Nrdc.org. Water Pollution Facts, Effects of Water Pollution, Clean Water Act | NRDC [Internet]. 2016 [cited 26 February 2016]. Available from: http://www.nrdc.org/water/

(9) Epa.gov. Recycling Basics | Reduce, Reuse, Recycle | US EPA [Internet]. 2016 [cited 26 February 2016]. Available from: http://www.epa.gov/recycle/recycling-basics

(10)   Toulmin C. Urban Agriculture: Diverse Activities and Benefits for City Society. Edited by C. Pearson, S. Pilgrim and J. Pretty. London: Earthscan (2010), pp. 126, £65.00. ISBN 978-1-84971-124-1. Experimental Agriculture. 2011;47(03):576.

(11)   Afdc.energy.gov. Alternative Fuels Data Center: Benefits and Considerations of Electricity as a Vehicle Fuel [Internet]. 2016 [cited 26 February 2016]. Available from: http://www.afdc.energy.gov/fuels/electricity_benefits.html

(12)   Lee I. Health Benefits of Walking. Medicine & Science in Sports & Exercise. 2007;39(Supplement):56.

(13)   Pollan M. The omnivore's dilemma. New York: Penguin; 2007.

(14)   Kids.usa.gov. Environment and Nature Jobs | Grades 6-8 | Kids.gov [Internet]. 2016 [cited 26 February 2016]. Available from: https://kids.usa.gov/teens/science/environment/environment-jobs/index.shtml

About the Author

Grace has a B.S. in Sociology, a Master's in Religious Education, and an Ed. S. in Global Training and Development. She is an associate professor of social studies, enjoys writing children's books, loves traveling, and lives with her family in the Northeast.

Made in the USA
Charleston, SC
04 September 2016